CARING FOR ROUGH GREEN SNAKE

UNDERSTANDING THE CHARMS AND CHALLENGES OF KEEPING A ROUGH GREEN SNAKE AS A PET AND INSIGHTS INTO THEIR BEHAVIOR AND CARE

DR MORRIS HART

Copyright© 2024 **DR MORRIS HART**

All rights reserved. No part or part of this book or publication may be reproduced, stored, or transferred in any form by electronic, mechanical, recording, or other retrieval system without written permission from the publisher

Table of Contents

INTRODUCTION: ROUGH GREEN SNAKES AS PETS 5

CHAPTER 1 11

GETTING STARTED: CREATING THE PERFECT HABITAT 11

CHAPTER 2 18

UNDERSTANDING THE BEHAVIOR OF ROUGH GREEN SNAKES.......18

CHAPTER 3 25

FEEDING A ROUGH GREEN SNAKE: TIPS AND TRICKS 25

CHAPTER 4 33

HANDLING AND SOCIALIZING: DEVELOPING TRUST WITH YOUR PET 33

CHAPTER 5 42

HEALTH AND WELLNESS: HOW TO KEEP YOUR ROUGH GREEN SNAKE HAPPY AND HEALTHY 42

CHAPTER 6 52

Troubleshooting Common Issues: Solutions for Harmony in Relationships 52

CHAPTER 7 62

Enrichment Activities to Stimulate Your Rough Green Snake's Mind and Body 62

CHAPTER 8 71

Conservation Awareness: promoting the preservation of wild rough green snake habitats 71

CHAPTER 9 82

Frequently Asked Questions (FAQs) about Rough Green Snakes as Pets.. 82

CHAPTER 10 89

Conclusion: The Happiness of Having a Rough Green Snake as a Pet 89

Introduction: Rough Green Snakes as Pets

Rough green snakes (Opheodrys aestivus), with their thin bodies and brilliant green hue, are fascinating reptiles that have grown in favor as pets among reptile aficionados. In this detailed book, we'll dig into the intriguing world of rough green snakes as pets, including their natural history, habitat requirements, food demands, and behavior, as well as practical tips on how to care for these magnificent creatures in captivity.

- Natural History

Rough green snakes, native to the southeastern United States, are arboreal reptiles that live in forested regions, particularly near water sources such as streams, swamps, and marshes. They are non-venomous and use their exceptional climbing ability and camouflage to avoid predators and search for prey.

Rough green snakes are mostly diurnal, which means they are most active during the day, and they frequently bask in the sun to regulate their temperature. Their food is mostly composed of small insects, spiders, and other invertebrates, making them useful for pest control in their natural habitat.

- Habitat requirements

When considering rough green snakes as pets, it is critical to mimic their natural environment as nearly as possible in a captive setting. Because these snakes are primarily arboreal and spend a lot of time investigating elevated locations, they require a vertically oriented terrarium or enclosure with plenty of climbing chances.

The enclosure should be supplied with branches, vines, and other structures for climbing and perching, as well as vegetation to provide cover and protection. Live plants, such as pothos or ficus, can assist create a lush and naturalistic habitat while also increasing the humidity levels in the enclosure.

Maintaining correct humidity is critical for the health and well-being of rough green snakes, as they require relatively high humidity levels to avoid dehydration and enable normal shedding. It is advised that humidity levels be monitored using a hygrometer and misted on a regular basis.

- Dietary Needs

In captivity, rough green snakes should be fed a diet that closely reflects their wild prey. Small crickets, fruit flies, moths, and other insects of the appropriate size are suitable dietary sources. A diversified diet is vital for maintaining nutritional balance and preventing dietary deficits.

Feeder insects should be gut-loaded with nutritious feeds and coated with calcium and vitamin supplements to ensure that the snake receives necessary nutrients. Adult rough green snakes are normally fed every 5-7 days, however younger snakes may need to be fed more frequently.

- Behavior

Understanding the behavior of rough green snakes is critical for providing proper care and enrichment in captivity. While these snakes are naturally docile and shy, they can grow stressed if not given appropriate hiding places and opportunities to flee imagined threats.

Rough green snakes are recognized for their delicate movements and arboreal lifestyle, typically employing their slender bodies to easily penetrate foliage and branches. They are great climbers and may spend a lot of time exploring the vertical space within their habitat.

Despite their mild nature, rough green snakes can become defensive if frightened or cornered. Handling should be maintained to a minimal and done with caution to avoid placing undue stress on the snake. When handling rough green snakes, it is critical to correctly support their bodies and minimize rapid movements.

Rough green snakes can be profitable and captivating pets for reptile enthusiasts who are prepared to provide them the specific care and attention they demand. Understanding their natural history, habitat requirements, food needs, and temperament allows pet owners to establish an environment in which these gorgeous snakes can survive in captivity. With proper care and stimulation, rough green snakes can provide years of enjoyment and curiosity to their owners.

Chapter 1

Getting Started: Creating the Perfect Habitat

Providing an optimum environment for a rough green snake (Opheodrys aestivus) is critical to ensuring its health, well-being, and overall quality of life in captivity. In this detailed tutorial, we'll look at the important components of creating the ideal environment for your rough green snake, from selecting the right enclosure to providing appropriate lighting, warmth, and furniture.

Choosing an Enclosure

The first step in preparing your rough green snake's environment is to choose an adequate enclosure. For a single adult rough green snake, a vertically oriented glass terrarium or reptile habitat measuring at least 18 inches long, 12 inches wide, and 24 inches tall is ideal.

Larger enclosures allow for more climbing and exploring, which is beneficial to both the snake's physical and mental health.

It is critical to select an enclosure with safe locking mechanisms to prevent escape, as rough green snakes are capable climbers who may seek to explore outside of their habitat if given the opportunity. In addition, the enclosure should include a screen or mesh top to provide for proper ventilation while keeping the snake from escaping.

Substrate

The substrate, or bedding, is essential for maintaining optimum humidity levels and providing a comfortable surface for your rough green snake to slither and burrow on. Organic potting soil, coconut fiber, and reptile-specific substrates suited for tropical species are all suitable substrate options.

Avoid using coarse or abrasive substrates since they can cause skin irritation and abrasions in rough green snakes. To avoid waste and bacteria growth, keep the substrate clean and replace it on a regular basis.

Decor and furnishings

To enrich and stimulate your rough green snake, furnish the enclosure with branches, vines, and other climbing structures. Live or artificial plants can be used to create a naturalistic setting while also providing cover and hiding locations for the snake.

When selecting decor and furnishings for the enclosure, choose materials that are safe and non-toxic to reptiles. Avoid using sharp or rough-edged things that could damage the snake, and make sure all items are securely attached to keep them from dropping and inflicting injury.

Heating & Lighting

Rough green snakes are ectothermic reptiles that regulate their body temperature using external heat sources, so maintaining proper temperatures is crucial for their health and well-being. To allow the snake to thermoregulate effectively, give a combination of ambient heat and basking places.

A heat source, like as an under-tank heating pad or ceramic heat emitter, can give ambient warmth in the enclosure, whereas a basking light or heat lamp can produce a localized heat source for the snake to bask in. Temperature gradients should be established within the enclosure, with one end warmer (about 85-90°F) for basking and the other somewhat colder (approximately 75-80°F) to allow the snake to regulate its body temperature as necessary.

In addition to heating, good illumination is essential for rough green snakes. Full-spectrum UVB lighting should be used to replicate natural sunlight and promote vitamin D production and calcium metabolism in the snake's body. UVB bulbs should be updated every 6-12 months to keep them effective.

Humidity & Water

Maintaining optimum humidity levels is critical for the health and well-being of rough green snakes, which require fairly high humidity to avoid dehydration and enable normal shedding. Humidity levels in the cage should be kept between 50 and 70%, with occasional sprinkling of the substrate and foliage to boost humidity as needed.

A shallow water dish should be placed in the enclosure so that the snake can drink and soak as needed. To ensure cleanliness, change the water on a regular basis, and make sure the dish is large enough for the snake to comfortably immerse itself.

Temperature and Humidity Monitoring

The temperature and humidity conditions of your rough green snake's enclosure must be monitored on a regular basis to ensure their health and well-being. To precisely monitor temperature and humidity levels, use digital thermometers and hygrometers, with probes set at both the warm and cool ends of the cage to guarantee a correct gradient.

Adjustments to the heating, lighting, and misting schedules may be required to maintain optimal conditions within the enclosure, particularly during seasonal changes or fluctuations in ambient room temperature and humidity.

Following these principles and providing a well-designed habitat that fulfills your rough green snake's physical, behavioral, and environmental demands will allow you to provide a safe, comfortable, and enriching environment for your pet to thrive in captivity. With appropriate care and attention to detail, your rough green snake can live a happy and healthy life in its new home for many years to come.

Chapter 2

Understanding the behavior of rough green snakes

Rough green snakes (Opheodrys aestivus) are intriguing reptiles with distinct characteristics that have captivated both researchers and reptile enthusiasts. In this detailed examination, we'll delve into the complexities of rough green snake behavior, including its inherent instincts, communication methods, social interactions, and responses to environmental cues.

Natural instincts.

Rough green snakes, like many other reptiles, have natural characteristics that help them survive in the wild. These instincts control many elements of animal behavior, such as foraging, movement, defense, and reproduction.

One of the most distinguishing characteristics of rough green snakes is their arboreal nature. As expert climbers, they spend a lot of time in trees and bushes, using their slender bodies and strong grip to easily traverse through branches and foliage. This arboreal existence allows them to evade ground predators and have access to a wide variety of prey items.

Rough green snakes have another essential instinct: they can efficiently hide themselves in their surroundings. Their vivid green colors and thin bodies help them to blend in perfectly with the forest's foliage and plants, offering superb concealment against both predators and prey.

Communication Methods

While rough green snakes do not use vocal communication like other animals, they do have a variety of nonverbal communication mechanisms for conveying information to conspecifics and other creatures in their environment. One such way is to use body language and posture.

When attacked or alarmed, rough green snakes may take defensive positions to deter prospective predators. This can include flattening their bodies to seem larger, hissing or vibrating their tails, and attacking defensively if necessary. When rough green snakes feel safe and secure, they may exhibit relaxed body language, such as coiling or contentedly resting.

Rough green snakes communicate with other snakes and animals in their surroundings using chemical cues as well as body language. They have a specialized organ called the Jacobson's organ in the roof of their mouth that detects and interprets chemical signals from airborne and surface fragrance particles. This capacity allows them to find possible partners, detect prey, and traverse their surroundings effectively.

Social Interactions

While rough green snakes are mostly solitary, they may interact with conspecifics during the breeding season or when confronting other snakes in their region. Rough green snakes normally breed through courtship rituals such as male combat and mate selection, followed by copulation and egg laying by the female.

Male rough green snakes may fight with rival males to gain supremacy and access to breeding females. Combat behavior usually consists of tangling and pushing against one other with their bodies, occasionally followed by attempts to pin the opponent to the ground. The winner of these encounters is granted mating privileges with receptive females in the area.

During the breeding season, female rough green snakes may engage in territorial behavior, defending their preferred nesting places from other females and potential predators. After mating, females will search for suitable sites to lay their eggs, such as hidden cracks or foliage near water sources.

Responses to Environmental stimuli

Rough green snakes are extremely sensitive to environmental stimuli and can display a variety of behaviors in reaction to changes in their surroundings. Temperature, humidity, light levels, and other conditions can all influence their activity, feeding habits, and reproductive cycles.

Temperature is one of the most important environmental factors for rough green snakes. Ectothermic reptiles use external heat sources to regulate their body temperature and metabolic activity. Optimal temperature gradients in their environment enable them to thermoregulate successfully, traveling between warmer and cooler locations as needed to regulate their internal temperature.

Humidity is another important environmental component for rough green snakes, especially during the shedding period. High humidity levels weaken the snake's old skin, allowing it to shed in a single piece without difficulties. Inadequate humidity can cause retained shed, requiring care to prevent harm or infection.

Lighting conditions also influence the behavior of rough green snakes, which rely on natural sunshine to regulate their circadian rhythms and behavioral patterns. UVB lighting promotes vitamin D production and calcium metabolism in their bodies, which are critical for overall health and bone formation.

By learning more about rough green snake behavior, we can better appreciate the intricate and interesting lives of these attractive reptiles. From their basic instincts and communication methods to their social interactions and responses to environmental cues, rough green snakes demonstrate a diverse range of behaviors that indicate their adaptability and resilience in their natural environment. As responsible caretakers, we must consider these behavioral qualities when caring for rough green snakes in captivity, ensuring that their physical, mental, and emotional well-being is maintained to the greatest standards.

Chapter 3

Feeding a Rough Green Snake: Tips and Tricks

Feeding is an important part of care for rough green snakes (Opheodrys aestivus) in captivity because it directly affects their health, growth, and overall well-being. In this comprehensive guide, we'll look at the dietary requirements of rough green snakes, explain appropriate food sources, provide feeding suggestions and procedures, and address frequent feeding-related concerns that snake owners may face.

Dietary Needs

Rough green snakes are carnivorous reptiles that eat small insects, spiders, and other invertebrates in the wild. When kept in captivity, it's critical to imitate their natural diet as nearly as possible to ensure they get the nutrients they need for good health and development.

A diversified diet is essential for providing the nutritional requirements of rough green snakes. Crickets, flies, moths, grasshoppers, and small roaches are all good feeder insects. It is critical to provide a diverse range of prey items in order to provide a balanced diet and prevent nutritional deficits.

In addition to live feeder insects, some rough green snakes may eat pre-killed or frozen-thawed food. However, live prey is often favored because it enriches the snake's diet and promotes its natural hunting activities.

Feeding Schedule

Rough green snakes' feeding frequency varies according to their age, size, and metabolic rate. As a general rule, adult rough green snakes should be fed every 5-7 days, but younger snakes may need to be fed more frequently to assist growth and development.

It is critical to check your snake's physical condition and change the feeding plan accordingly. Underweight or rapidly growing snakes may require more frequent feedings, but overweight snakes may require a reduced feeding schedule to avoid obesity.

Feeding should take place during the afternoon, when rough green snakes are most active and alert. Offering food in the evening or at night may cause the snake to refuse to eat, as they are usually less active at these times.

Feeder insects

When choosing feeder insects for your rough green snake, make sure to select high-quality prey that is free of pesticides, parasites, and other toxins. Before feeding the snake, the feeder insects should be gut-loaded with healthy foods such leafy greens, fruits, and commercial insect gut-loaders.

Additionally, feeder insects should be appropriate for the snake's age and size. Feeder insects that are too large may constitute a choking threat or create digestive problems, whereas prey items that are too little may lack appropriate nourishment.

When arranging your snake's diet, you should also consider the nutritional value of feeder insects. Some insects, such as crickets and mealworms, contain more fat than others and should be consumed in moderation to avoid obesity and other health issues.

Supplementation

To ensure that rough green snakes get all of the important vitamins and minerals they require, dust feeder insects with calcium and vitamin supplements before feeding them to the snakes. Commercial reptile calcium supplements containing vitamin D3 are widely available and can be administered to feeder insects by lightly dusting them with the powder before feeding.

Supplementation should be used cautiously, as excess vitamins and minerals might be hazardous to the snake. Follow the manufacturer's dose and frequency of supplementing guidelines, and provide a well-balanced meal rich in prey to assist meet the snake's nutritional requirements.

Feeding Techniques

Feeding rough green snakes is a simple operation, but there are several strategies that can help encourage reluctant eaters or solve special feeding concerns. Here are some suggestions for feeding your rough green snake:

When offering feeder insects to your snake, use feeding tongs to hold the prey item and simulate movement. This can help to activate the snake's natural hunting instincts, encouraging it to strike and swallow the prey.

Live Prey: When possible, provide live feeder insects to your rough green snake. Live prey enriches the snake's diet and encourages natural hunting habits, helping to keep it active and interested.

Temperature: Before providing feeder insects to your snake, make sure they are at the optimum temperature. Insects that are too cold may be slow and less enticing to the snake, but insects that are too warm may become stressed and aggressive.

Feeding Enclosure: Some snake owners choose to feed their snakes in a separate feeding enclosure to reduce the potential of substrate ingestion and contamination. To feed the snake outside of its main enclosure, use a small, escape-proof container with smooth sides.

Monitor Feeding Response: Keep an eye on your snake's feeding response and change your feeding procedures as appropriate. Some snakes may prefer to hunt from elevated perches, whereas others may prefer to ambush prey from hidden locations.

Regurgitation: If your snake regurgitates its food quickly after eating, it could be due to poor care, handling stress, or other underlying health problems. Allow the snake to rest and recover before attempting to feed again, and speak with a veterinarian if regurgitation continues.

Feeding rough green snakes necessitates careful consideration of their nutritional requirements, feeding schedule, and feeding practices to ensure they receive the nourishment they require for good health and growth. Snake owners can help their rough green snakes survive in captivity by feeding them a variety of live feeder insects, supplementing as needed, and using effective feeding strategies. Monitoring the snake's feeding response and changing husbandry procedures as appropriate can also help address feeding concerns and maintain the snake's long-term health and nutrition.

Chapter 4

Handling and Socializing: Developing Trust with Your Pet

Handling and socializing are critical parts of keeping rough green snakes (Opheodrys aestivus) in captivity. Snake owners may form bonds with their pets and guarantee their physical and emotional well-being by instilling trust and good associations through careful handling and regular engagement. In this comprehensive tutorial, we'll go over the fundamentals of handling and socializing rough green snakes, as well as advice for safe and effective handling techniques and strategies for developing trust and confidence in your pet snake.

Understanding Snake Behavior

Before getting into handling and socialization strategies, it's critical to understand rough green snakes' natural behavior and how they perceive and respond to human interaction. Rough green snakes, like many other reptiles, prefer to be alone and avoid social interactions with humans and other animals. Handling may be perceived as a stressful or threatening activity, particularly if not conducted with care and respect.

When dealing with rough green snakes, it's critical to be aware of their body language and communication clues. Stress or discomfort can be indicated by hissing, fast breathing, defensive postures, and attempts to flee or conceal. It is critical to follow these cues and avoid handling the snake if it appears irritated or defensive.

Establishing Trust

Building trust with your rough green snake takes time, persistence, and positive reinforcement. Allow the snake to adjust to its new surroundings and build a sense of security in its container before attempting to handle it. Spend time near the enclosure, chatting softly to the snake and offering food, so that it associates your presence with happy events.

Begin with brief handling sessions and progressively increase the duration as the snake gets more comfortable. Handle the snake with care and confidence, supporting its body and letting it move at its own rate. Avoid rapid movements or prolonged handling sessions, as these might create stress and pain for the snake.

Handling Techniques

When dealing with rough green snakes, it is critical to employ suitable practices to ensure the safety and well-being of both the snake and the handler. Here are some guidelines for safe and efficient handling:

Approach the snake carefully and gently, allowing it to see and smell you before attempting to touch it. Avoid making rapid movements or loud noises that could shock or frighten the snake.

Support the Body: When lifting up the snake, use both hands to support it, one for the front half and the other for the back. Avoid gripping or holding the snake too firmly, since this may cause tension and discomfort.

Avoid the Head: When handling tough green snakes, it is advisable to avoid touching the head or neck if possible. Snakes may interpret handling around their heads as a danger and respond defensively.

Be Patient: Handling sessions should be done at the snake's own pace, allowing it to explore and move freely in your hands. Be patient and compassionate, and don't force the snake to interact if it appears stressed or unwilling.

Keep an eye out for signs of stress, such as the snake's body language and behavior during handling sessions. If the snake shows signs of stress or discomfort, such as hissing, rapid breathing, or attempted escape, gently return it to its container and try again later.

Wash Your Hands Before and After: To avoid the spread of bacteria and illness, you must wash your hands before and after handling your snake. This protects you and your pet from hazardous diseases.

Socializing Techniques

While rough green snakes may not seek social connections with humans in the same way that dogs or cats do, regular handling and positive reinforcement can help your pet snake gain trust and confidence over time. Here are some ways to socialize rough green snakes:

Handling your snake on a daily basis will help it develop acclimated to human interaction while also reducing fear or stress connected with handling. Short, frequent handling sessions are more beneficial than longer, less frequent ones.

Positive Reinforcement: Provide food treats or rewards during handling sessions to instill a positive relationship with human touch. This can teach the snake to identify handling with happy experiences, reducing fear and anxiety.

Gentle Touch: Gently stroke the snake's body during handling sessions to assist it develop acclimated to human contact. Avoid touching sensitive portions of the snake, such as its head or tail, as this may lead it to become protective.

Respect Boundaries: While handling is necessary for socializing rough green snakes, it is also critical to respect the snake's limits and preferences. If the snake appears stressed or uncomfortable, give it some room and try again later.

Encourage exploration by allowing the snake to explore its environment and interact with various objects and textures during handling sessions. This can pique the snake's interest and enrich its diet.

Building Confidence

To instill confidence in your rough green snake, create a safe and supportive atmosphere in which it feels secure and comfortable. Provide hiding places, climbing chances, and other enrichment activities within the enclosure to stimulate natural behaviours and mental stimulation.

Additionally, avoid sudden movements or loud noises that may startle or frighten the snake, and handle it carefully and respectfully to gradually create trust and confidence.

Handling and socializing rough green snakes needs patience, consistency, and a thorough grasp of their natural behavior and communication signs. Snake owners who approach handling sessions with care and respect can foster trust and positive associations with their pets, ultimately strengthening the link between human and snake. With good handling techniques, regular socialization, and a supportive environment, rough green snakes can live in captivity and continue to delight and fascinate their owners for years.

Chapter 5

Health and Wellness: How to Keep Your Rough Green Snake Happy and Healthy

Maintaining the health and well-being of your rough green snake (Opheodrys aestivus) is critical to ensuring a long and pleasant life in captivity. This thorough book will cover the most important areas of snake health and wellness, such as habitat management, feeding and nutrition, hydration, handling, and common health issues. You can help your rough green snake thrive and live a healthy life by providing proper care and taking proactive actions.

Habitat Management

Creating an appropriate habitat is the foundation of snake health and fitness. A well-designed cage should mimic the snake's natural environment while providing opportunities for activity, enrichment, and thermoregulation. Here are some important aspects to consider when managing your rough green snake's habitat.

Enclosure Size: Choose an enclosure with plenty of room for your snake to move and explore. For an adult rough green snake, a vertically oriented enclosure measuring at least 18 inches long, 12 inches wide, and 24 inches tall is recommended.

Substrate: Select a substrate that promotes cleanliness while allowing for natural digging and nesting habits. Coconut fiber, cypress mulch, and organic potting soil are all suitable substrate possibilities. Avoid abrasive or dusty substrates, as these might irritate the snake's skin and respiratory tract.

Maintain a temperature differential within the enclosure so that your snake can thermoregulate successfully. At one end of the enclosure, provide a basking zone with temperatures ranging from 85 to 90 degrees Fahrenheit, and at the other end, provide a colder area with temps ranging from 75 to 80 degrees.

Rough green snakes require somewhat high humidity levels to avoid dehydration and promote normal shedding. Maintain a humidity level of 50-70% by spraying the cage on a regular basis and providing a shallow water dish for drinking and soaking.

Enrichment: Encourage your snake's natural habits by giving enrichment possibilities such as climbing branches, hiding places, and environmental stimuli. Rotate and vary enrichment materials on a regular basis to keep things interesting and stimulate discovery.

Feeding and Nutrition

Your rough green snake's health and vitality depend on a well-balanced diet. In the wild, rough green snakes eat mostly insects and invertebrates. When feeding your snake in captivity, it's critical to provide a range of prey items to maintain nutritional balance. Here are some suggestions for eating and nutrition:

Feeding Schedule: Feed adult rough green snakes every 5-7 days, and juvenile snakes every 3-5 days, to promote growth and development. Feed your snake according to its age, size, and metabolic rate.

Provide a variety of feeder insects, including crickets, flies, moths, and small roaches. To ensure your snake gets the nutrients it needs, gut-load feeder insects with nutritious foods and dust them with calcium and vitamin supplements before feeding.

Supplementation: To avoid nutritional deficits, add calcium and vitamin supplements to your snake's food. Dust feeder insects with supplements before feeding, or keep a calcium supplement in a separate dish for your snake to eat as needed.

Monitor Body Condition: Check your snake's body condition on a regular basis to ensure it has a healthy weight and appetite. Adjust feeding frequency and portion quantities as needed to avoid obesity or malnutrition.

Hydration

Proper hydration is essential for snake health and avoiding dehydration. Rough green snakes get water from both their dietary and drinking sources. Make sure your snake has access to clean, fresh water at all times by placing a shallow water dish in its habitat. Monitor water levels on a regular basis and refill as needed to avoid dehydration.

In addition to drinking water, rough green snakes may benefit from the occasional dip in shallow water dishes. Soaking hydrates the snake's skin and aids in the shedding process, ensuring that the old skin sheds in one piece without difficulties.

Handling and Socialization

Handling and socializing are critical parts of snake care that foster trust and confidence in your rough green snake. When handling your snake, proceed with caution and confidence, supporting its body and allowing it to move at its own pace. Avoid making rapid movements or loud noises that could shock or frighten the snake.

Regular handling sessions can assist your snake become desensitized to human contact and lessen the stress associated with handling. Begin with short, gentle handling sessions and gradually increase the length and frequency as your snake gets more comfortable.

While handling is good for socializing, remember to respect your snake's boundaries and preferences. If the snake appears anxious or defensive, give it some room and try again later. Not all snakes appreciate handling, so emphasize your snake's comfort and well-being.

Common Health Issues:

Despite your best efforts to give good care, rough green snakes may develop health problems from time to time. Being alert and proactive in monitoring your snake's health can aid in the early detection and resolution of potential issues. Here are some common health issues to look for:

Respiratory infections can occur as a result of inappropriate husbandry, such as low temperature or humidity conditions. Symptoms may include wheezing, open-mouth breathing, nasal discharge, and fatigue. If you feel your snake has a respiratory illness, contact a veterinarian right once.

Parasites: Internal and external parasites can infect rough green snakes, particularly those taken in the wild. Weight loss, low appetite, tiredness, and the presence of visible parasites in the feces or on the skin are all possible symptoms. Consult a veterinarian for proper diagnosis and treatment.

Dehydration: Rough green snakes are prone to dehydration, particularly when humidity levels are low or water sources are scarce. Dehydration symptoms include sunken eyes, wrinkled skin, lethargy, and a decreased appetite. Ensure that your snake always has access to clean, fresh water, and monitor humidity levels on a regular basis.

Shedding Issues: Rough green snakes lose their skin as they grow. Inadequate humidity or nutritional inadequacies can cause shedding issues, such as retained or incomplete sheds. Maintain appropriate humidity levels and provide regular soaking opportunities to aid in the shedding process.

Maintaining the health and wellness of your rough green snake necessitates careful attention to its environment, diet, hydration, handling, and overall health. By providing a proper environment, adequate nutrition, regular hydration, careful handling, and proactive health monitoring, you can help your snake have a long, healthy, and joyful life in captivity. Remember to seek veterinarian help right away if you observe any signs of disease or health problems, as early intervention is critical to successful treatment and recovery. With proper care and attention, your rough green snake will thrive and bring delight and curiosity into your life for many years.

Chapter 6

Troubleshooting Common Issues: Solutions for Harmony in Relationships

Developing a harmonious relationship with your rough green snake (Opheodrys aestivus) entails discussing and resolving common concerns that may occur during the course of snake ownership. From handling difficulties to behavioral issues, proactive troubleshooting can help you and your pet snake form a pleasant and fulfilling association. In this article, we'll look at some of the most common challenges that snake owners have and offer practical strategies to solve them.

Handling Difficulties

Handling difficulties are a regular challenge for snake owners, particularly those who are new to snake ownership or have minimal knowledge with reptiles. When handled inappropriately or with insufficient trust and confidence, snakes, even rough green snakes, may exhibit defensive actions such as hissing, striking, or attempting to flee. Here are some tips for dealing with difficulties:

Start Slowly: If your snake is new to handling, allow it to adjust to your presence near its enclosure. Spend time sitting near the enclosure and talking softly to the snake to assist it become accustomed to your scent and voice.

utilize Proper Technique: When handling your snake, utilize proper handling techniques to reduce stress and protect the safety of both you and the snake. Support the snake's body firmly but gently, and avoid any rapid movements or harsh handling that may shock or alarm it.

Be Patient: Establishing trust and confidence with your snake requires time and patience. Begin with short handling sessions and progressively extend the time as the snake becomes more comfortable. If the snake becomes stressed or defensive, gently return it to its enclosure and attempt again later.

Positive Reinforcement: Provide food treats or rewards during handling sessions to instill a positive relationship with human touch. This can help the snake link handling with happy experiences, eventually reducing fear or anxiety.

Respect Boundaries: When handling your snake, you must respect its boundaries and preferences. If the snake appears anxious or defensive, give it some room and try again later. Not all snakes appreciate handling, so emphasize your snake's comfort and well-being.

Food Refusal

Feeding refusal is another typical problem for snake owners, especially with recently acquired or agitated snakes. Rough green snakes may refuse to feed for a variety of causes, including stress, disease, environmental problems, or a general lack of appetite. Here are some techniques for addressing food refusal:

Examine Husbandry: Make sure your snake's enclosure is properly set up with adequate temperatures, humidity levels, and hiding areas. Improper snake husbandry might stress them out and cause them to refuse food.

Experiment with several varieties of feeder insects to determine what your snake prefers. Some snakes prefer live prey, while others will consume pre-killed or frozen-thawed prey items. Providing a range of prey possibilities can help to increase the snake's appetite.

Feed at Different Times: Feeding your snake during its natural active hours, which are usually during the day, may enhance the likelihood of adoption. Avoid feeding the snake when it is brumating or shedding, as its appetite may be limited.

Limit Handling: Reduce handling and other causes of stress around feeding time to make your snake feel more at ease and relaxed. Avoid upsetting the snake right before or after feeding because it may interrupt the feeding response.

Monitor Weight: Keep track of your snake's weight and body condition to ensure that it has a healthy appetite. If feeding rejection persists, or if your snake exhibits signs of weight loss or other health problems, see a veterinarian for further evaluation and advice.

Shedding Problems

Shedding issues, such as retained or partial sheds, can occur in rough green snakes due to low humidity conditions, nutritional deficits, or other causes. Shedding difficulties can cause discomfort, irritation, and an increased risk of skin infections. Here are some suggestions for addressing shedding issues:

Maintain Proper Humidity: To assist healthy shedding, keep the humidity levels in your snake's enclosure between 50 and 70%. Maintaining proper humidity levels can be achieved by spraying the enclosure on a regular basis and providing a shallow water dish for soaking.

Provide Soaking Opportunities: If your snake is having problems shedding, place shallow water bowls or dampened paper towels for soaking. Soaking can soften old skin and aid in the shedding process.

Check Husbandry: Examine your snake's husbandry habits, such as temperature, humidity, and substrate, to find any probable shedding causes. Make changes as needed to establish the best environment for shedding.

Provide Rough Surfaces: Rough green snakes may benefit from rough surfaces in their enclosure, such as rough branches or rocks, to help them shed. These surfaces might help the snake shed old skin as it crawls around.

Avoid Handling During Shedding: Reduce handling and other sources of stress during shedding times so that your snake may concentrate on the shedding process. Avoid upsetting the snake or attempting to remove shed skin prematurely, as this might result in harm or incomplete sheds.

Health Monitoring

Regular health monitoring is critical for detecting and treating potential health problems in rough green snakes. Here are a few crucial indicators to look for:

Body Condition: Check your snake's body condition on a regular basis to ensure it is at a healthy weight and muscular tone. Symptoms of being underweight or overweight include visible ribs or spine, excessive fat deposits, and changes in appetite or activity level.

Respiratory Health: Look for symptoms of a respiratory illness, such as wheezing, open-mouth breathing, nasal discharge, or difficulty breathing. Environmental stresses, such as low temperatures or humidity, or pathogen exposure, can all lead to respiratory illnesses.

Skin Health: Examine your snake's skin for symptoms of dehydration, retained shed, and skin blemishes. Healthy snake skin should be smooth and free of wounds, abrasions, and discolouration. Retained shed can develop when humidity levels are too low or the snake is suffering from another health problem.

Monitor your snake's appetite and digestive condition to ensure it is eating and passing stool properly. Appetite changes, vomiting, regurgitation, or diarrhea may signal underlying health problems requiring veterinarian care.

Behavioral Changes: Pay close attention to any changes in your snake's behavior, activity level, or temperament. Behavioral changes may indicate stress, disease, or other health issues that necessitate further investigation.

Maintaining a healthy connection with your rough green snake requires proactive troubleshooting and attentive care to handle frequent difficulties while also promoting the snake's health and well-being. Understanding your snake's behavior, providing an appropriate habitat, providing proper feeding and water, and routinely evaluating its health can all help guarantee that your pet snake lives a long and enjoyable life. Remember to seek veterinarian help right away if you observe any signs of disease or health problems, as early intervention is critical to successful treatment and recovery. With patience, consistency, and proper care, you and your rough green snake can have a satisfying and enriching connection for many years.

Chapter 7

Enrichment Activities to Stimulate Your Rough Green Snake's Mind and Body

Enrichment activities are essential in keeping rough green snakes (Opheodrys aestivus) cognitively and physically occupied in captivity. Rough green snakes, being interested and active reptiles, benefit from a variety of environmental stimuli that mirror their native habitat and promote natural behaviors. In this comprehensive guide, we'll look at the benefits of enrichment for rough green snakes, go over numerous enrichment activities and tools, and give you practical advice on how to incorporate enrichment into your snake's daily routine.

Understanding Enrichment

Enrichment is the providing of exciting and engaging experiences that improve the physical, mental, and emotional well-being of captive animals. Enrichment activities are critical for rough green snakes because they avoid boredom, reduce stress, and encourage natural behaviors like exploration, hunting, and climbing.

Enrichment can take many forms, including environmental enrichment (providing a dynamic and stimulating habitat), social enrichment (interactions with conspecifics or other animals), sensory enrichment (stimulating the senses through sight, sound, smell, and touch), and cognitive enrichment.

Advantages of Enrichment for Rough Green Snakes
Enrichment activities provide several physical and psychological benefits to rough green snakes. Some of the primary benefits are:

Preventing Boredom: Enrichment activities provide diversity and interest, preventing boredom and creating a more gratifying environment for the snake.

Engaging in natural behaviors and activities reduces stress and anxiety in caged snakes, resulting in increased overall well-being.

Physical Health: Enrichment activities promote physical activity, exploration, and exercise, which improves muscular tone, flexibility, and cardiovascular health.

Stimulating Mental Acuity: Enrichment activities test the snakes' cognitive powers and problem-solving abilities, keeping them alert and engaged.

Enhancing Natural Behaviors: Enrichment activities allow rough green snakes to engage in natural behaviors including climbing, burrowing, foraging, and basking, which promote species-specific behaviors and instincts.

Enrichment Activities for Rough Green Snakes

Snake owners can utilize a variety of enrichment activities and instruments to stimulate and improve the quality of life for their rough green snakes. Below are some examples of enrichment activities for rough green snakes:

Climbing Structures: Rough green snakes are arboreal creatures that prefer climbing and investigating vertical areas. Provide your snake with a variety of climbing structures to investigate, including as branches, vines, and driftwood.

Hiding locations: Provide several hiding locations within the enclosure to ensure your snake's safety and solitude.

To provide a safe haven for your snake to relax and hide, use realistic hides such as cork bark, hollow logs, or artificial caves.

Rough green snakes may like to burrow in loose substrates like coconut fiber, cypress mulch, or organic potting soil. Create a deep layer of substrate for your snake to burrow and dig in, replicating its natural burrowing behavior.

Sensory Stimuli: Use sensory enrichment materials like textured surfaces, aromatic plants, or aural stimuli to pique your snake's interest. Allow your snake to explore and interact with items of varying textures and fragrances.

Feeding Enrichment: Use feeding enrichment tactics to help your snake develop natural hunting and foraging behaviors. To make feeding more hard and exciting, offer live prey such as crickets, flies, or small roaches, as well as puzzle feeders and hiding locations.

Environmental Changes: Rotate and modify enrichment items on a regular basis to avoid boredom and keep your snake's habitat interesting. Change the arrangement of the enclosure, add new decorations, or introduce unique items to keep your snake interested and engaged.

Handling and Socialization: Rough green snakes benefit from regular handling and socialization sessions, which allow them to connect with their human caretakers and explore new settings beyond their enclosure.

outside Enclosures: If the weather and climate circumstances allow, consider giving outside enclosures or supervised outdoor time for your snake to get some natural sunlight, fresh air, and stimulation.

Tips to Incorporate Enrichment

When introducing enrichment activities into your rough green snake's daily routine, remember the following success tips:

Observe Your Snake: Pay attention to your snake's behavior and preferences to figure out which enrichment activities they like the most. Some snakes prefer to climb and explore, while others enjoy burrowing or hunting.

Begin slowly: Introduce enrichment activities and monitor your snake's reaction. Allow your snake to explore and engage with new objects at its own pace, and be patient while it adapts to new stimuli.

Safety first: Make sure that all enrichment objects and activities are safe and appropriate for your snake's size, age, and behavior. Avoid using things with sharp edges, hazardous compounds, or small bits that could be swallowed or constitute a choking risk.

Supervise Outdoor Time: If you provide outdoor enclosures or supervised outdoor time, make sure the location is safe and devoid of potential threats like predators, harmful plants, or harsh weather conditions. To avoid escapes or injury, keep your snake under strict supervision.

Rotate and Vary Enrichment: To avoid boredom and keep your snake's environment interesting, rotate and vary enrichment items on a regular basis. Introduce new things, modify the cage arrangement, and offer novel stimuli to keep your snake interested and curious.

Monitor Behavior: Keep track of your snake's behavior and well-being on a frequent basis to ensure that enrichment activities are effective. Keep an eye out for signs of tension, boredom, or discomfort, and make any necessary alterations to offer your snake with a good and enriching environment.

Enrichment activities are critical for maintaining the health, happiness, and well-being of rough green snakes in captivity. Snake owners may help their pets grow and live meaningful lives by creating a fascinating and engaging environment that resembles their natural habitat and fosters natural activities. There are numerous ways to improve your snake's daily routine, including climbing structures, hiding locations, sensory stimuli, feeding enrichment, and outdoor habitats. By studying your snake's behavior, experimenting with various enrichment activities, and putting safety and well-being first, you can create a rich and pleasant environment for your rough green snake to enjoy for many years.

Chapter 8

Conservation Awareness: promoting the preservation of wild rough green snake habitats

Conservation awareness is critical in encouraging the preservation of wild rough green snake (Opheodrys aestivus) habitats and guaranteeing the species' long-term survival in its native environment. As human activities continue to have an impact on natural ecosystems, it is critical that individuals, communities, and organizations take action to protect and conserve the habitats that rough green snakes and other wildlife depend on. In this comprehensive book, we'll look at the value of conservation awareness, the risks to wild rough green snake populations, and practical ways for supporting habitat preservation and biodiversity protection.

Understanding the Rough Green Snake Habitat

Rough green snakes are native to North America, where they live in a variety of woodland and grassland environments. These slender, arboreal snakes are commonly found in densely vegetated habitats such as woods, woodlands, meadows, and marshes. Rough green snakes prefer riparian areas near water sources, where they can find abundant prey and ideal breeding locations.

Rough green snakes rely on a healthy ecosystem that includes ample vegetation, prey species, and adequate microclimates for thermoregulation. They spend a lot of time climbing and hunting in the understory plants, where their slender bodies and exceptional camouflage help them blend in with the environment.

Threats to Rough Green Snake Habitats.

Despite their versatility, rough green snakes face a number of risks to their habitat and survival, the majority of which are caused by human activity and habitat degradation. Some of the primary risks to rough green snake habitats are:

Habitat Loss and Fragmentation: The conversion of natural habitats for agriculture, urban development, and infrastructure has resulted in the loss and fragmentation of rough green snake habitat. Fragmentation breaks down connection between habitat areas, separating populations and decreasing genetic diversity.

Deforestation: Clearing forests and woodlands for timber extraction, agriculture, and urbanization removes essential rough green snake habitat, limiting the availability of suitable nesting locations, foraging grounds, and shelter.

Pollution: Pollution from agricultural runoff, industrial operations, and urbanization can deteriorate water quality and pollute rough green snake habitats, rendering them uninhabitable and endangering the health of aquatic and terrestrial ecosystems.

Invasive plant species like Japanese honeysuckle (Lonicera japonica) and Chinese privet (Ligustrum sinense) can outcompete native vegetation and disrupt ecosystem dynamics, lowering habitat quality for rough green snakes and other native species.

Climate Change: As temperatures and precipitation patterns change, rough green snakes and other species will experience variations in habitat appropriateness and distribution. Rising temperatures can disrupt hibernation and breeding cycles, while extreme weather can destroy habitats and kill animals.

Importance of Conservation Awareness

Conservation awareness is critical in combating the threats to rough green snake habitats and encouraging the preservation of natural ecosystems. Individuals and communities may help safeguard rough green snakes and other species for future generations by promoting awareness of the importance of biodiversity conservation, habitat preservation, and sustainable land management techniques. Conservation awareness is crucial for several reasons, including:

Conservation awareness programs serve to educate the public about rough green snakes' ecological relevance and role in preserving healthy ecosystems. Individuals can support conservation efforts by becoming more aware of the hazards to rough green snake ecosystems.

Building Support for Conservation Initiatives: Conservation awareness campaigns help to increase public support for conservation policies, financing programs, and habitat restoration projects. Conservation organizations and government agencies can use public support to execute effective conservation plans to maintain rough green snake habitats and biodiversity.

Promoting Sustainable Practices: Conservation awareness motivates individuals, businesses, and policymakers to use sustainable land management practices to reduce habitat destruction, pollution, and other threats to rough green snake habitat. Conservation efforts can help offset the effects of human activity on natural ecosystems by encouraging sustainable agriculture, forestry, and urban design.

Conservation awareness enables local communities to take ownership of conservation activities and contribute to habitat restoration, monitoring, and stewardship operations. Conservation organizations can achieve conservation goals by engaging stakeholders and developing community partnerships.

Conservation awareness develops a sense of connection to nature, as well as respect for wildlife's beauty and diversity. Conservation initiatives can foster a better knowledge of the significance of rough green snake habitats and the need of preserving them for future generations by encouraging people to engage with nature and enjoy the wonders of wild areas.

Promoting Habitat Preservation

There are numerous approaches for individuals, communities, and organizations to promote habitat preservation and conservation awareness in order to maintain rough green snake ecosystems. Here are some practical ways to promote habitat preservation:

Education and Outreach: Plan educational programs, workshops, and outreach events to raise awareness about rough green snake habitats, biodiversity protection, and the necessity of protecting natural ecosystems. Schools, community groups, landowners, and policymakers are examples of different audiences to target.

Support habitat restoration programs aimed at restoring damaged or fragmented rough green snake habitats through reforestation, native plants, invasive species removal, and habitat improvement. Volunteer with local conservation organizations or participate in citizen science projects to help with habitat restoration efforts.

Advocacy and Policy Support: Promote laws and regulations that conserve rough green snake habitats and encourage sustainable land use practices. Support legislation that preserves key habitat areas, creates protected areas, and limits activities that endanger wildlife and ecosystems.

Community Engagement: Work with local communities and stakeholders to instill a sense of stewardship and shared responsibility for habitat preservation. Encourage community participation in conservation planning, land management decisions, and habitat monitoring efforts.

Partnerships and Collaboration: Work with government agencies, conservation organizations, academic institutions, and other stakeholders to pool resources, knowledge, and take collective action to preserve habitat. Form collaborations with landowners, businesses, and industry stakeholders to promote sustainable land management and conservation easements.

Public Awareness Campaigns: Launch public awareness campaigns emphasizing the importance of rough green snake habitats, highlighting success stories of habitat preservation and restoration, and encouraging individuals to support conservation efforts. Reach out to a diverse audience and promote conservation messages using multimedia platforms, social media, and traditional outreach approaches.

Conservation awareness is critical to promoting the preservation of wild rough green snake habitats and guaranteeing the species' long-term survival in its native environment. Individuals, communities, and organizations can help protect rough green snakes and other wildlife for future generations by raising awareness about the threats to their habitats, educating the public about the importance of biodiversity conservation, and promoting habitat preservation and sustainable land management practices. We can protect rough green snake habitats and our natural world's rich biodiversity by collaborating on education, advocacy, habitat restoration, community participation, and public awareness initiatives.

Chapter 9

Frequently Asked Questions (FAQs) about Rough Green Snakes as Pets

Rough green snakes (Opheodrys aestivus) are fascinating and unusual reptiles that make excellent pets for reptile aficionados. Owning a rough green snake, like any other pet, raises a number of issues and considerations. In this FAQ guide, we'll answer some of the most frequently asked questions regarding owning rough green snakes as pets, including care requirements, behavior, health, and more.

1. What are the fundamental care requirements for rough-green snakes?

Rough green snakes require less maintenance than other reptile species. They require a correctly sized container with enough ventilation, a suitable substrate for burrowing, a temperature gradient of 75-85°F, and moderate humidity levels of 50-70%. Rough green snakes also require a diversified diet of small insects and invertebrates, fresh water for drinking and soaking, and frequent handling to maintain socialization.

2. What size enclosure do rough green snakes require?

Rough green snakes are arboreal and need a vertical habitat to support their climbing activity. A 20-gallon tall terrarium (about 18 inches long, 18 inches wide, and 24 inches tall) will accommodate one adult rough green snake. As rough green snakes are good climbers, make sure the enclosure has a sturdy cover to prevent escapes.

3. What do rough-green snakes eat?

In the wild, rough green snakes eat small insects and invertebrates like crickets, flies, moths, and roaches. In captivity, they can be fed a range of live or pre-killed prey, such as crickets of the appropriate size, flightless fruit flies, and mealworms. A diversified diet is vital for maintaining nutritional balance.

4. How frequently should I feed my rough green snake?

Adult rough green snakes should be fed every 5-7 days, however juvenile snakes may need more frequent feedings (every 3-5 days) to assist their growth and development. Adjust your snake's feeding frequency based on its age, size, and metabolic rate, and keep an eye on its bodily condition to avoid overfeeding or underfeeding.

5. Do rough green snakes require specialized lighting?

Rough green snakes do not require UVB lighting, unlike several other reptile species. However, maintaining a natural day-night cycle with a consistent light-dark schedule will enhance your snake's circadian rhythms and foster natural behaviors. A basic incandescent or LED bulb can offer heat and light to your snake's habitat.

6. How do I manage my rough green snake?

When handling your rough green snake, proceed with caution and confidence, supporting its body securely yet softly. Begin with short handling sessions and gradually increase the length and frequency as your snake becomes more comfortable. Avoid making rapid movements or loud noises that could shock or frighten the snake, and always wash your hands before and after handling to prevent the transmission of bacteria.

7. Are rough green snakes suitable for beginners?

Rough green snakes are often regarded suitable for novice reptile keepers because to their gentle demeanor, low care requirements, and tiny size. They do, however, have unique environmental requirements, like as humidity and temperature, which necessitate careful consideration. As with any pet, proper study and planning are essential for providing maximum care for rough green snakes.

8. How do I know whether my rough green snake is healthy?

A healthy rough green snake has clean eyes, smooth and vivid skin, a robust appetite, and sheds on a regular basis. Monitor your snake's behavior and activity level for any signs of disease or stress, such as lethargy, loss of appetite, respiratory issues, or unusual behavior. Regular health check-ups with a reptile veterinarian are also recommended to keep your snake in good condition.

9. Are rough green snakes suitable pets for children?

Rough green snakes can be wonderful pets for children who are responsible and grown enough to handle them carefully and delicately. However, adult supervision is always required when handling reptiles, particularly for small children. Educate children on correct snake handling skills and the significance of respecting the snake's boundaries and requirements.

10. How long can rough green snakes live in captivity?

With proper care, rough green snakes can survive for 6-8 years in captivity, while some individuals may live longer with proper husbandry and veterinary treatment. Providing an appropriate home, a balanced diet, regular handling, and proactive health monitoring can all help your rough green snake have a long and healthy life in captivity.

Keeping rough green snakes as pets may be a fulfilling experience for reptile enthusiasts of all skill levels. Understanding your rough green snake's fundamental care requirements, eating habits, handling practices, and health issues will allow you to provide a fascinating and interesting environment in captivity. If you have any additional questions or concerns about caring for your rough green snake, speak with a competent reptile doctor or experienced reptile keeper for personalized advice and help.

Chapter 10

Conclusion: The Happiness of Having a Rough Green Snake as a Pet

Keeping a rough green snake (Opheodrys aestivus) as a pet may be an enjoyable and rewarding experience for reptile enthusiasts of all skill levels. These thin, arboreal snakes are noted for their mild disposition, striking appearance, and low maintenance requirements, making them ideal selections for reptile keepers looking for a distinctive and engaging pet. In this final section, we'll discuss the joys and benefits of keeping a rough green snake as a pet, as well as the obligations and considerations that come with reptile ownership.

1. Unique and Fascinating Pets

Rough green snakes are unlike any other pet, with their bright green coloration, slender bodies, and beautiful movements. Their arboreal nature and climbing ability make them intriguing to observe as they move around their enclosures and explore their surroundings. Their quiet demeanor and gentle disposition make them perfect pets for reptile lovers who appreciate studying and engaging with their pets in a peaceful environment.

2. Educational Opportunities

Keeping a rough green snake as a pet offers excellent educational opportunities for people of all ages. Learning about rough green snakes' natural history, behavior, and habitat requirements can help pet owners get a better understanding of ecology, biodiversity, and the interconnection of all living organisms. Educating others about rough green snakes and their role in the ecosystem can also assist to raise awareness about the value of animal conservation and habitat preservation.

3. Therapeutic Benefits

Interacting with pets, particularly rough green snakes, can provide therapeutic advantages to pet owners by lowering stress, anxiety, and sadness. A pet snake's relaxing presence, along with the habit of caring for and observing them, can provide a sense of camaraderie, purpose, and emotional comfort. Individuals with special needs or disabilities may find that interacting with a calm and reliable pet snake provides a unique sort of sensory stimulation and social engagement.

4. Connection with Nature

Keeping a rough green snake as a pet creates a stronger connection with nature and helps pet owners to appreciate wildlife's beauty and diversity. Pet snakes remind us of our collective obligation to maintain and preserve natural ecosystems for future generations by allowing us to see into the fascinating lives of wild animals. Pet ownership can foster feelings of wonder, curiosity, and reverence for nature.

5. Responsible Pet Ownership.

Owning a rough green snake entails many obligations and concerns to maintain the health, safety, and well-being of both the pet and its owner. Pet owners must provide a comfortable environment with appropriate temperature, humidity, and lighting, as well as a healthy nutrition and regular veterinarian care. They must also handle their pet snake with caution and responsibility, taking steps to avoid escapes, injuries, and stress.

6. Advocate for Conservation

People who keep rough green snakes as pets can become champions for wildlife protection and habitat preservation. Pet owners have a unique opportunity to raise awareness about the risks to rough green snake populations in the wild, as well as to encourage appropriate land management practices that safeguard natural ecosystems. They can also help conservation organizations, participate in citizen scientific projects, and fight for legislation that benefit wildlife and ecosystems.

7. Creating a bond

Developing a bond with a rough green snake may be a highly satisfying experience that benefits both the pet and the owner. Regular handling, engagement, and care can help pet owners create a trusting and mutually beneficial relationship with their snake, promoting feelings of friendship, connection, and affection. The link between a pet snake and its owner is beyond language and species borders, providing a unique type of companionship and understanding.

8. Inspiring Others

Keeping a rough green snake as a pet can teach others to value and respect reptiles and wildlife. Pet owners can foster interest, empathy, and admiration for rough green snakes by sharing their experiences, knowledge, and passion for them. Pet owners may inspire individuals of all ages to care about reptiles and conservation through educational outreach, social media, and community engagement.

Finally, having a rough green snake as a pet provides a plethora of joys, benefits, and chances for pet owners to connect with nature, promote conservation, and form meaningful relationships with their animals. From the distinct beauty and behavior of rough green snakes to the educational and therapeutic benefits of pet ownership, the pleasures of owning a rough green snake reach well beyond its enclosure. Pet owners can make a positive difference in the lives of these fascinating creatures and promote a greater respect for the natural world by accepting the responsibilities of pet ownership, campaigning for wildlife protection, and sharing their enthusiasm for rough green snakes with others. As stewards of the land and its inhabitants, we must continue to respect and safeguard the valuable biodiversity that enhances our lives and nourishes our planet.

www.ingramcontent.com/pod-product-compliance
Lightning Source LLC
Chambersburg PA
CBHW050231230526
45470CB00005B/1905